Animals have always been a part of Wild Mike's life since he was little. From then on, he has continued to pursue his passion to care for some of the most amazing creatures on this planet as a zookeeper. Over the years, he has yearned for a way to share the beauty and importance zoological institutions hold for everyone. His simple photography hobby has turned into a new remarkable craft that has allowed Wild Mike to share an amazing look into the world of animals he loves so much.

THE *Wanderlust* KEEPER

Written & Photographed by
WILD MIKE MILLER

Austin Macauley Publishers™
LONDON • CAMBRIDGE • NEW YORK • SHARJAH

Copyright © Wild Mike Miller (2021)

All rights reserved. No part of this publication may be reproduced, distributed, or transmitted in any form or by any means, including photocopying, recording, or other electronic or mechanical methods, without the prior written permission of the publisher, except in the case of brief quotations embodied in critical reviews and certain other non-commercial uses permitted by copyright law. For permission requests, write to the publisher.

Any person who commits any unauthorized act in relation to this publication may be liable to criminal prosecution and civil claims for damages.

Ordering Information
Quantity sales: Special discounts are available on quantity purchases by corporations, associations, and others. For details, contact the publisher at the address below.

Publisher's Cataloging-in-Publication data
Miller, Wild Mike
The Wanderlust Keeper

ISBN 9781647500641 (Paperback)
ISBN 9781647500634 (Hardback)
ISBN 9781647500658 (ePub e-book)

Library of Congress Control Number: 2020919794

www.austinmacauley.com/us

First Published (2021)
Austin Macauley Publishers LLC
40 Wall Street, 33rd Floor, Suite 3302
New York, NY 10005
USA

mail-usa@austinmacauley.com
+1 (646) 5125767

I would like to take the opportunity to properly thank the people who made this project become a reality.

First to my family and close friends, who always believed in me and stuck by my side as I traveled around the globe. They were happy to share in my excitement over the littlest things while never squashing my enthusiasm. Next, a giant show of appreciation is deserved to all my dear friends at Jersey Smokes Cigar Shop in Jackson, NJ. Not only was the shop kind enough to host a fundraising event in my honor to assist in my travels, but all the loyal customers came out to show their undying support and interest in my work. All my family and friends are beyond amazing for dealing with a crazy kid like me. I love you all and hope that you realize I will always be there for you as you were there for me.

Secondly, this definitely would not have been possible without the photographic permissions granted to me by all the zoological institutions featured in this book. They welcomed me to their facilities with open arms and always maintained an open line of communication if I had any questions or problems. It was a privilege and honor to share their uniqueness with you. They allowed me to explore a passion for photography that will always have me being eternally grateful. To learn more about these institutions along with all their efforts in conservation, please visit their websites, which will all be listed in the bibliography.

Lastly, I want to extend a word of perpetual gratitude to all my fellow animal care professionals, conservationists, researchers, and scientists that have dedicated their lives to wildlife and the success of nature. Without all of you, who knows where we would find ourselves. Keep up the truly excellent work!

Foreword

It's hardly surprising that the first time Mike Miller and I shook hands, it involved fantastic creatures and his camera. We have one major thing in common, a profound love for animals. For me, turtles and tortoises, in particular, had secured themselves a paramount placement in my life at the early age of just five years, and by the time I met Mike, they had literally become my life.

Lucky for me, my superhuman wife, Casey, also came to share her life with these creatures and together we built our very own breeding and rescue facility known as Garden State Tortoise, from the ground up. Casey then became responsible for Mike and I beginning a solid friendship; thanks to her days working at Six Flags Safari as a zookeeper where Mike is employed as a warden. Our friendship was inevitable, of course, as Mike's driving desire to witness and photograph wildlife would bring him to our place, where he could capture photos of species he'd never seen before. His positivity is contagious and the passion he has for animals is revealed within seconds of beginning a conversation. Fully captivated by the wonder and beauty of the specimens in our collection, Mike became the camera, securing moments both tranquil and magnificent.

My work with turtles and tortoises has spanned nearly three decades, and while this stretch of time has certainly featured some noteworthy highlights, there was something peculiar and special about the photographs Mike took of our animals. In addition, his overall goal in publishing books that display the boundless importance of zoos and other facilities working themselves to the bone to responsibly care for and propagate the planet's threatened and endangered species is needed more than ever. In today's world, social media has crushed any other educational platform by forcing every speck of surfacing information down our throats. We can hardly breathe as we are fed story after story and opinion after opinion. But that's just it, most of what we are fed is merely a chaotic collection of opinions. Sadly, zoos have taken on the brunt of negative opinions explicitly because our visions and true understandings have been distorted by heavy, relentless influence from news feeds. On our planet earth, can we really find a faithfully untouched place where man has yet to leave his mark? We may find ourselves scrutinizing the thought of what is truly wild in our current world, but as species accelerate toward extinction, animal facilities fight to preserve them. In the end, these places may be responsible for all that remains as so many species slip by, becoming only the ghosts of what they once were in the natural world.

Throughout this masterpiece, the author's extensive travel accompanied by thrilling camera work takes us to some remarkable scenes. These safeguarded creatures represent the future for their species as the colors of the big picture we humans have painted continue to run. Mike Miller's "The Wanderlust Keeper" is an engaging personal account that encompasses both his journey to places far and near, as well as a reflection of the undeniable positivity that these facilities have created for animals.

So, I ask you, the reader, to open your mind and heart to the astounding mortals that are still with us as a consequence of the deeply devoted caregivers in animal facilities throughout our world. Enjoy this wild ride.

Chris Leone
Owner, Garden State Tortoise
Director of Animal Husbandry, TheTurtleRoom

First Light

Conservation Status In The Wild

Not Evaluated
Data Deficient
Least Concern
Near Threatened
Vulnerable
Endangered
Critically Endangered
Extinct In The Wild
Extinct

Introduction

Daybreak
North Atlantic Ocean

The word "Wanderlust" has a few simple definitions that make it seem uneventful and lacking feeling. One such example is "the strong desire to travel," or according to Merriam-Webster, a "strong longing for or impulse toward wandering." That is indeed a clear and direct explanation. However, to me, over the years it has been a state of mind that has kept me full of life and wonder. When you pick up a dictionary and you locate the word "wanderlust", it should say: "To have a thirst to push boundaries geographically as well as mentally in order to experience all the world may have to offer." Everyone has different reasons for why they would want to travel. Some want to explore natural wonders, while others may travel to experience a different culture or investigate a piece of history. In my case, I aim to see how zoological institutions around the world are doing their part in preserving the planet's most precious creatures. Over the course of a year, I have wandered to seven institutions on three continents, over 35,000 miles and filled a camera with around 3,000 photos. I chose these specific facilities because they each have a uniqueness that put themselves among top in the animal care field. They range from having a rich history, innovative designs, stunning overall size, and phenomenal conservation programs. I had discovered that those highlights were just the tip of the iceberg.

At this very moment I am completing the final journey of this book. As the sun sets behind the Glasshouse Mountains, the air is consumed with sounds of countless insects and frogs just starting their day. Lorikeet and Cockatoo calls start to fade as they settle into their evening roosts. Hard to believe that I am sitting in the middle of it all, finding it to be the most tranquil of spots I have been to in nearly a decade. I am fully aware how lucky I am to experience all of this. This has taken hard work, planning, saving, and a helping hand from those who have always wanted me to succeed. To all of those who have helped me and continue to show me support, I am extremely thankful.

With this undertaking I hope to show the everlasting beauty these creatures possess through my photography. I am positive that is exactly what you will find as you course through these pages. All of this wandering had a purpose. To learn all I could and to share all the knowledge gained, so together we can all do our part to conserve the treasures of this earth. So sit back and enjoy. It was an amazing journey and thank you for joining me.

Odysea Aquarium™

The first leg of this journey starts in a spot where I am surrounded by over two million gallons of water and sandy desert, simultaneously. Now I know many of you may have guessed an oasis and, in a way, you are right. In Scottsdale, Arizona sits the first jewel of this wandering keeper's exploits, OdySea Aquarium™. It stands as the largest aquarium in the Southwest region of the United States and in the top five across the rest of the country. Its design was part of a ten year vision that came to life after just eighteen months of construction once the project was initiated. Contrary to how you may have seen buildings erected in the past, this project was planned for the amazing future residents from the get-go. Being built from the inside out allowed the exhibits to take top priority and full attention. As you explore this vessel, anyone can affirm that the animal's well-being and guest experience take precedence.

The aquarium leads the guests on an interactive adventure by means of immersing them in various experiences from start to finish. When guiding yourself past seventy exhibits, you come face to face with nearly 370 species, totaling around 6,000 individuals. One minute you may be kneeling to get a close look at eight inch tall Lined Seahorses and later find an eight foot long California Sea Lion gliding past you.

As someone who is naturally curious with a thirst for knowledge, I was in my element. They take one on a voyage that faces a drop of water as it falls from the clouds into the many lakes or rivers around the world and inevitably into the oceans. You never find one to be puzzled at an exhibit because the signage surrounding you educates on another level. No matter what your age, you become compelled to take the chances of involvement seen around almost every corner. Going face to face with a massive sturgeon at one of the touch pools or getting wrapped up in the Octopus keeper talk. Hell, there even was a large water table, which lets you take the hands on approach to learning how dams and gradients affect water flow.

Innovation and individuality are two words that filled my eyes as I came across the aquarium's fish globes, Seatrek™, and the world's first rotating Exhibit Theater. Before you've even had the ticket for entry scanned, there is a surprise exhibit ten feet above your head. A specially designed group of nine fish globes that seemingly float in the foyer. They weigh 1,000 pounds apiece and each are equipped with their very own Life Support System (LSS) for the species held within. Thanks to some clever engineering, the couple hundred feet of pipes that sustain the globes are out of sight, allowing the guest to take in the species beauty without distraction.

As you travel on, there comes an opportunity to be fully immersed from head to toe in the Seatrek™ 255,000 gallon experience. The Seatrek™ experience can be found at over 50 other locations around the planet, but the one found at OdySea AquariumTM stands above the rest. It is the only location which showcases the Indo-Pacific ocean region and is completely housed indoors. After receiving your sterilized wet suit, safety briefing, and nautical diving helmet, the dive team guides you around fifty-seven beautiful species. With such an amazing connection built between the public and sea life it is easy to see why there have been thirty fantastic marriage proposals and counting at this site.

Then I decided to take a seat in the world's first rotating exhibit theater. As I settled in and the lights dimmed, I was overcome with an excited curiosity of what I was going to see. Was I about to see the inner workings, like in the movie Jurassic Park? No, it was even better. It was like being on a submarine with nothing but you and the creatures of the ocean deep. You are guided through four exhibits, where some very important messages are passed on to the people when it comes to sea turtles and sharks.

The aquarium is home to several sea turtles that were found injured and cannot return to the wild. Luckily, the amazing facilities and medical staff here were able to nurse these guys back to health so they can live a full life and educate those who could help their wild cousins. As the world gets smaller, sea turtles are just one of many species that have suffered. Amputations of fins are common due to abandoned crab pots or discarded nets that easily get wrapped around their appendages. In other instances, when the turtles surface to breath they can suffer damage to the shell from propellers and boats. In this exhibit you can see the result of such an injury. It can result into a condition known as "bubble-butt." Comical as it sounds, this is indeed a serious situation. The damage caused will create a pocket under the shell, thus affecting the animal's ability to properly swim. Without that, escaping predators or finding food would be near impossible. As you will see, the veterinary staff is able to counteract the effects by surgically affixing weights to the shell. As the turtle grows, the weights will be adjusted accordingly.

Next up was the shark exhibit where you are in the presence of several species including Sand Tiger, Lemon, and Nurse. Regrettably, years of misunderstanding these species combined with a high-demand for shark fin soup has led to nearly 70 million sharks meeting an unfortunate end every year. The truth is sharks are immensely important to maintaining a balance in our oceans. As apex predators, they control the equilibrium of the whole food chain. By feeding on fish/squid/etc. that are weak or sick, they keep the diversity among other species varied and healthy. As seen before in other ecosystems, if the apex predators disappear, then a ripple effect will find its way causing vast damage in our oceans.

OdySea AquariumTM not only educates about conservation, but they also are very active participants at home and abroad. Members of the aquarium team, along with a crew of volunteers, regularly organize a visit to the Coon Bluffs to clean up the area. Trash left behind by absent minded visitors can, with greatest of ease, find their way in the waterways and then into the food chain. For example, a plastic bag floating in the water bares a remarkable resemblance to a jellyfish which just so happens to be the prey of choice for a sea turtle. Another event I regrettably missed was the free Annual Conservation Expo hosted by the OdySea Aquarium Foundation™. Right outside the facility, over 40 organizations are brought together, which bring many different ideas for all members of the public to get involved and do their part to help out. Such organizations include Arizona Game and Fish Department, as well as, the Arizona Humane Society.

On an international level, the aquarium partners with a non-profit organization located over 9,000 miles away in South Africa, SanccobTM. Created in the 1960s, SanccobTM has primarily worked to combat falling seabird populations. This is done by diligent rehab/release programs, in addition to hand rearing abandoned chicks (over 8,000 since 2001) of different species, including the endangered African Penguin. Since the African Penguin is an endangered species of focus for SanccobTM, the aquarium has been able to use their penguin exhibit to help this conservation group. They will host up close educational presentations that build a connection between the public and these charismatic creatures. Secondly, they also have hosted fund raisers to monetarily help the charity by allowing the penguins to explore their creative side with finger painting, well in this case, webbed feet painting. The paintings were made available to the public with all funds raised going straight to SanccobTM. Just one of the many unique ways this facility has built a base in order to help another specie's future survival,

I will never forget hearing as a child that "we know more about outer space then we know about our oceans." Now, I have faith that the trend is shifting in the other direction. So without further delay, I offer you a look at the many species from around the world that I learned about at OdySea Aquarium™.

Floating Seas
Electric Yellow Cichlid
Status: Least Concern

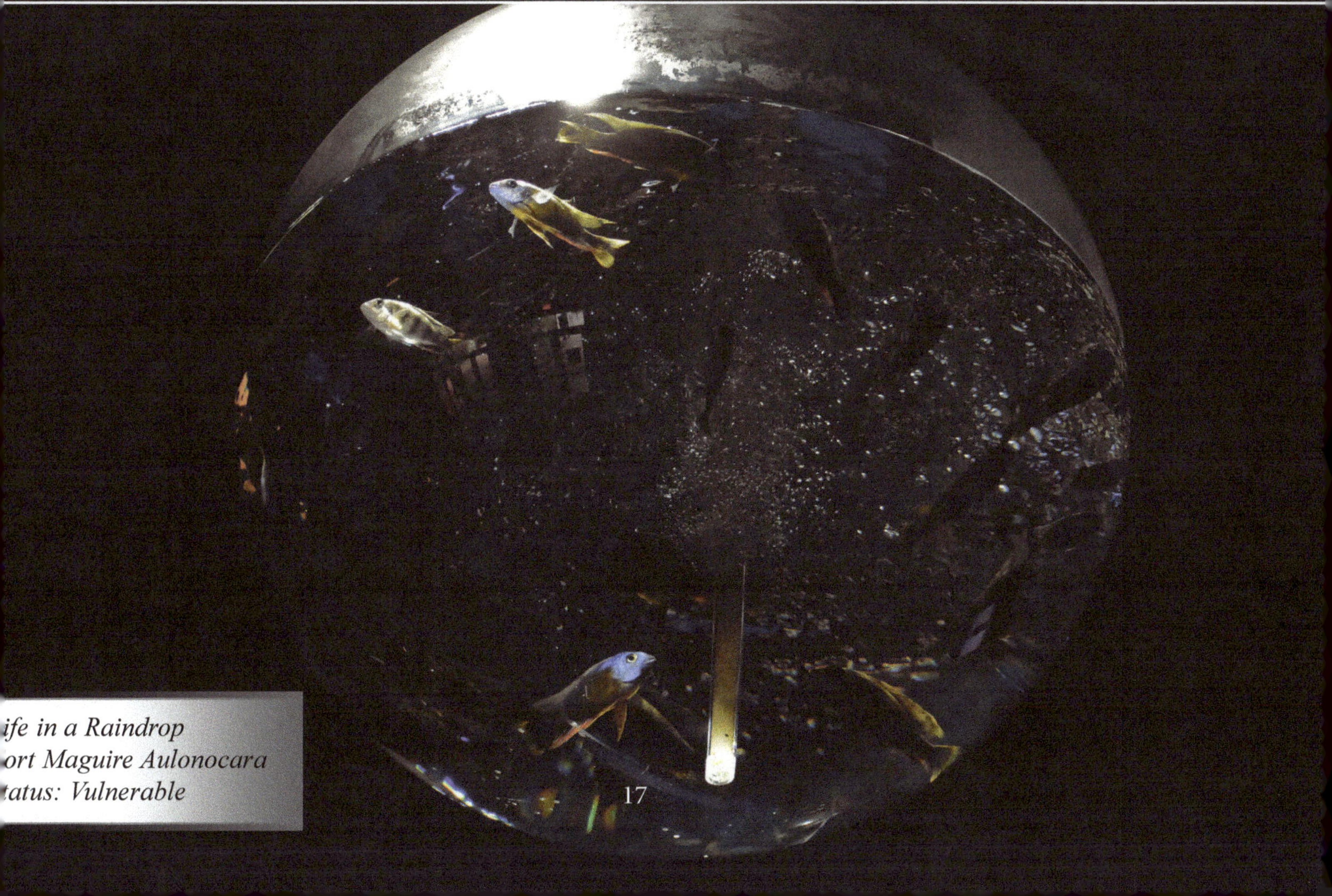

Life in a Raindrop
Fort Maguire Aulonocara
Status: Vulnerable

Spiked Illusion
Russian Sturgeon
Status: Critically Endangered

rimitive Face

Giant Among Minnows
Colorado Pikeminnow
Status: Vulnerable

Apex Youth
Leopard Shark
Status: Least Concern

Mirrored Giant
Arapaima
Status: Data Deficient

Slow Movers
Black Pacu
Status: Not Evaluated

Tri-Color Wonder
Keel-billed Toucan
Status: Least Concern

Sharp Cuddle Puddle
Morelet's Crocodile
Status: Least Concern

Gleaming Horde
San Francisco Piranha
Status: Not Evaluated

Sea Legs
Walking Batfish
Status: Least Concern

Morining Stretch
Gaint Pacific Octopus
Status: Least Concern

Herculean Suction

Solitary Invader
Lionfish
Status: Least Concern

Oddity of the sea
Giant Gourami
Status: Least Concern

Entangled Water horse
Lined Seahorse
Status: Vulnerable

Hidden Gaint
Goliath Grouper
Status: Critically Endangered

Conference Among the depths
Lemon Shark
Status: Near Threatened

Gliding Power House
Sand Tiger Shark
Status: Vulnerabl

Diving Angel
California Sea Lion
Status: Least Concern

Gliding Angel

Microscopic Adaptation
Brown Banded Bamboo Shark
Status: Near Threatened

Beauty of the Reef
Zebra Shark
Status: Vulnerable

Repaired at Sea
Loggerhead Sea Turtle
Status: Near Threatened

New Lease on Life

Life at the Top

North Carolina Zoo

The Gateway

    The one great thing about winter is that spring is not very far off. Trust me when I tell you that after working outside through the frigid temperatures for months that your mind is solely focused on a few things. You anticipate the arrival of longer days and a fresh start that brings better moods to all forms of life. I could think of no better spot to kick-start the return of spring then at the North Carolina Zoo. Just a stone's throw outside of Asheboro, North Carolina lays the world's largest natural habitat zoo. With over five miles of walking paths to cover, you truly are transcended into the wild. What I found so profound was how in a setting like this not only do you form a connection with nature, but you become part of it.

    Established in 1974, the North Carolina Zoo started with two Galapagos Tortoises and has now grown to sustain over 1,600 animals, 52,000 plants on about 500 developed acres. With such an immense platform to potentially inspire the public, this institution offers one thing that made me truly honored to be there. Whenever a North Carolina educational institution decides to visit on a trip, the students are welcomed to the zoo with free admission! Education is one of the most important values that we can pass on to the next generation. With over 100,000 North Carolina school children visiting each year, I have faith that many will be inspired to preserve our world so the amazing creatures among us can last for their kids to see. When children learn so many amazing things during a single experience, they are just itching to tell their family, friends, and neighbors all about it. All those great stories lead many to experience it for themselves, and in 2017 the zoo set a calendar year attendance record of over 860,000 guests.

These guests are all treated to animals from two amazing regions of this planet: North America and Africa. In between these areas I had plenty to experience. There was beautiful animal artwork, a Honeybee Garden, and in 2018 the zoo hosted an amazing "Birds in Flight" educational show. Two days was almost not enough time to take it all in, but you can believe I did not waste any time.

In the North America section, there is a hidden treasure that many people never realized used to be common around the Southeastern United States, the Red Wolf. This species used to roam widely in the region but habitat destruction and poaching nearly sent them to the realm of extinction. This facility has become a key member of the Red Wolf Recovery Program alongside several other institutions. With about 30 individuals left in the wild of the Carolinas, this project has become pivotal in the battle combating the specie's decline. Hosting the second largest population of Red Wolves in the world, North Carolina Zoo has been able to help increase the captive population to about 230 individuals in cooperative breeding programs among 40 other institutions. The zoo is also working hand in hand with land owners around eastern North Carolina to keep the territories of the last wild roaming wolves preserved and secure future spots for possible reintroductions.

As I continued toward the Africa section, I found what happens to be the best honey bee learning experience out of the 50 institutions I have visited. With the bees allowed to come and go as they please, you never know if you will have the chance to see them do what they do best. Lucky for me the queen and her subjects had just arrived to the zoo's Honey Bee Garden the day before. The hive is constructed so the public can watch the importantly busy lifestyle of this insect, while they remain pleasantly undisturbed. Surrounding the hive are vibrant educational boards and members of the North Carolina Honey Bee Association who volunteer to spread the word in regards to this phenomenal creature. We all know they help pollinate plants which grow into our food and other goods, but let me just put it into perspective how truly important they are. In a single year, the tireless efforts of bees help to yield $96 million worth of fruits/vegetables as well as $90 million worth of alfalfa, cotton, and soybeans. It goes to show just how much we rely on just one living thing.

As I still wander beneath the shaded pathways, I arrive at a breathtaking view point, the Watani Grasslands exhibit.

I am now faced with 40 open acres that can be seen from several possible viewpoints hosting a wide range of species like the African Elephant, Fringe-eared Oryx, and Southern White Rhinoceros. The serenity of the rolling hills and seeing the animals without a care in the world is enough to make you question if you are still in the United States or arriving on the Serengeti. All the care and design of this space is to show the public what could be lost forever if conservation is left to being an afterthought. The North Carolina Zoo fully embraces that importance and is involved in programs and research to protect habitats/species in eight different African countries, areas of the Caribbean, and throughout the state.

The exhibit that I connected with the most was the Western Lowland Gorilla troop. These great apes never cease to amaze me and are, by far, my favorite to photograph. From their sheer strength, anthropomorphic expressions, and social interactions, I could observe them all day without ever caring how much time has passed. The more I watched, the more I was honored to learn. This troop receives a phenomenal diet of restaurant quality fruit and vegetables throughout the day.
This allows them to continually express their natural foraging behavior, mirroring the same amount of time that their wild cousins spend doing the same activity. This is where I learned about the zoo's efforts to help the world's most endangered ape, the Cross River Gorilla.

On the border of Nigeria and Cameroon, there are only about 300 Cross River Gorillas left. North Carolina Zoo partners with another wildlife conservation organization to combat the biggest threat to this species, poaching. Working directly with the area rangers, they supply them with tools that are at the forefront of technology. Thus, allowing them to track gorilla movements while fishing out poacher traps/snares much more effectively. People in all walks of life can attest to the fact that without the right tools nothing can ever get done or change for the better.
I was truly astounded by the flawless incorporation of the North Carolina forest into the exhibits. With an additional 2,100 acres that the zoo rests on, we can only hope that there are plans in the future to grow this amazing institution. Even if that may not be the case, the power of education presented to the public here will undoubtedly inspire countless generations for years to come

Find the Seeker
Red Wolf
Status: Critically Endangered

Ghost of the East Woods

The Patrol

Future Leader

Queen of the Ice
Polar Bear
Status: Vulnerable

Glacier Giant

Cascading Quadruped
Grizzly Bear
Status: Least Concern

Handsome Forager

Resting Forest King
Cougar
Status: Least Concern

Artwork: "Black Dog"
Artist: Donna Dobberfuhl

Left: Honeycomb Tango
Above: Returning Scout

Western Honey Bee
Status: Data Deficient

Delayed Takeoff
Sun Conure
Status: Endangered

Rainbow in Flight

A Prey's View
Eurasian Eagle-owl
Status: Least Concern

Young King
King Vulture
Status: Least Concern

Nesting Queen
Victoria Crowned Pigeon
Status: Near Threatened

Artwork: "Nature's Recyclers"
Artist: Chris Gabriel

Raw Power
African Lion
Status: Vulnerable

Stone Pachyderm
Rock Hyrax
Status: Least Concern

Free Range
Plains Zebra
Status: Near Threatened

Wild Runner

Fringe Territory
Fringe-eared Oryx
Status: Vulnerabl

Ivory Mirage
African Elephant
Status: Vulnerable

Gargantuan Sunrise

Double Tuskers

The Last Bite
Western Lowland Gorilla
Status: Critically Endangered

Watchful Eye

Anthropomorphic

Learning All

Omaha's Henry Doorly Zoo and Aquarium

Opening Day

Now that the woes of winter are finally behind me, the spiritual travel bug within me kicked into high gear as I landed in Omaha, Nebraska. Now I know many of you are thinking, "Miller! Why are you going all the way there?! You are going to get lost in a corn field." Even though, with my luck, that may have been a real possibility. I always like to take necessary risks. When I researched what awaited me at Omaha's Henry Doorly Zoo and Aquarium, I could barely contain my excitement.

With about 960 different species and approximately 17,000 individuals, the zoo is among some of the largest in the world as it pertains to animal residents. As if that wasn't enough to have me virtually jumping over the aisles to disembark my plane, they are a facility that has created some truly marvelous exhibit designs with immersive concepts. These include America's largest indoor rainforest (Lied Jungle®), the world's largest indoor desert (Desert Dome), indoor swamp, and nocturnal exhibit (Eugene T. Mahoney Kingdom of the Night®). Omaha Zoo and Aquarium also constantly immerse the guest in education around every corner, from the exhibits mentioned above to the Hubbard Gorilla Valley. To cap off the entire experience, they are involved in some revolutionary conservation programs and practices.

As rainforests around the world vanish with incredible haste, Lied Jungle® takes you into the heart of precisely why conservation is so important. Every detail in this 123,000 sq. ft. space places you into life's most diverse ecosystem. Upon walking in the door, your senses are vaulted beyond what you can imagine. I explored every inch, from the forest floor to the top of the tree canopy. Modeling the rainforest from three parts of the world (South America, Africa, and Asia), combined with corresponding mixed species exhibits give every guest a front row seat to the wilds in this unexpected backyard. Even the plants were seamlessly incorporated throughout the guest pathways and exhibits properly from their endemic regions.

Such as the Coconut Palms of Southeast Asia to the Epiphytic Orchids of South America. This experience made me speechless, which some of my closest friends can attest is not an easy feat. It left me yearning for the chance to head to the closest rainforest, but that journey will have to wait its turn. For now, standing at the base of Omaha Zoo and Aquarium, fifty foot waterfall while a Baird's Tapir forages for fallen fruit from the Black Handed Spider Monkeys above is pretty darn close to my kind of perfect.

With so much to see and so little time, it was just a short walk to the next remarkable feature, the Desert Dome. You may be wondering what kind of building would encapsulate the world's largest indoor desert, swamp, and nocturnal exhibit.

Well, none would be as fitting unless it was the world's largest geodesic dome. Measuring in at 137 feet high and with a diameter of 230 feet, you really have to remember to blink as you gaze upon this marvel, and that is even before you get into the building!

When I was told I was about to step foot into the desert environments of the Namib, Sonoran, and Red Center of Australia, I was both intrigued and nervous. For I know those environments can be ruthlessly hot and it was very sunny outside of the glass dome. Of course, those moments of trepidation were quickly extinguished due to the engineering greatness that keeps all the guests and animal residents comfortable through the day. An amazingly intricate water cooling system works its magic overnight by generating ice. As the day commences, it begins to melt and is fed through special pipes traveling under guest pathways and other areas creating a comfortably average temperature of around 73 degrees Fahrenheit. Every tile of the dome was purposefully positioned or tinted to create maximum shade in the summer and allow the highest light input during the winter months. These optimal conditions allow for every species, from Klipspringer to the Rattlesnakes, a place to thrive. This presents a unique experience showing how such unforgiving environments have so many creatures that fulfill critical
niches in nature.

After exploring the world of desert life, it was time to go underground into Eugene T. Mahoney Kingdom of the Night®, which includes the swamp, and I could not believe my eyes. Literally, it was so perfectly made as a nocturnal environment, I had to wait a second for my eyes to adjust. Ha! See what I did there? Anyway, to create a pleasing understanding of nocturnal species special lifestyle, such as the quick footed Springhaas, light cycles are reversed. Thus, replicating the regular time of activity to the public who normally would be fast asleep while nature is in motion.

Special lights throughout allow you to see the animals and does not provide any discomfort to the residents. This kind of environment does present its challenges while trying to photograph without a flash, all while in extremely low light.
However, I never have backed down from a challenge before and now was not

the time to start. Going through the 42,000 sq. ft. nocturnal area you are transported into a beautiful cave setting with a couple thousand stalactites, 70 ft. high shaft, and a glowing cavernous lake. Once again, this facility hit the nail on the head when it comes to being immersed in an environment unlike any you may have been in before.

There is a flawless flow into this institution's take on the swamp atmosphere and does not disappoint. From the sounds and smells to the species surrounding the very piers you walk on. The bayous of the Southern United States have come to life. The most notable apex predator of the swamp, American Alligator, has always been a favorite to see. From their longevity as a species to their critical role in the environment, it is hard to not appreciate these creatures. You even had a chance to learn about a distinctive genetic circumstance, leucism, when you meet their Leucistic American Alligator. All it means that, in the case of this gator, it was born lacking pigmentation in its skin but otherwise is just like any other member of its species. This attribute can be seen in many different species, from birds all the way to giraffes. Not to be confused with albinism, which has the species born completely without pigment in the skin and eyes. After a much needed night of sleep it was time for me to explore the other two-thirds of this great institution.

My favorite learning experience of this trip came when I arrived to the Hubbard Gorilla Valley. Over the years I have seen many Western Lowland Gorilla exhibits, but I still found myself taking in new bits of knowledge and even more appreciation for this species. Omaha Zoo and Aquarium is partnered with other zoological facilities across five continents in a united effort to conserve this iconic species. In house they offer a detailed insight into the growth and development of these massive gentle giants with brilliant educational displays. Luckily, for me, the zoo had a young resident who will show guests just how fast they grow up. Every day it will explore, learn, and teach those who visit the zoo, all the beauty and importance of gorillas to nature. Another great visual learning opportunity presents itself through the halls. There you find replica skeletons of all the Great Apes, some lesser primates, and a human. Being able to have a 360 degree view of the same bones we all possess allows for a connection on a deeper level. Omaha's Henry Doorly Zoo and Aquarium are definitely doing their part to build a connection between its guests and nature.

Conservation efforts are not just restricted to out in the wild, but also hold some very important prospects behind the scenes in laboratory research. Omaha Zoo and Aquarium has worked hand in hand with The Reproductive Science

Department to perfect many techniques including AI (artificial insemination). One of the noteworthy fruits of their efforts led to the first Tigers born through AI. With every single scientific victory, you come to find advancements which will inevitably lead to creating a chance for our friends in the wild to thrive.

One such possibility has led to these collaborating scientists to create a method to remove disease generating organisms from reproductive material trialed on native species, such as a buffalo or various antelope species in South Africa. Not only would this give one less risk to wild animal populations, but could possibly assist humans in the future. However, all of this work would mean nothing if we don't take an extra step to care for the countless plant life all around us.

As we learn in some of our earliest science classes, plants are an integral part of our lives. They form daily diets, medicines, and even supply materials for basic shelter. It has been determined that when a single plant species vanishes, anywhere from about ten to thirty other organism can follow the same unfortunate fate. Omaha's Henry Doorly Zoo and Aquarium has been working with many other outlets, including U.S. Fish and Wildlife, to try and conserve all manner of plants. By having plants endemic to the same areas as the animals around the zoo, they are showing the dependency that exists between them. They have successfully led restoration projects of threatened plant species, as well as perfecting methods in the realm of cryogenically storing seeds.

All the efforts from the immersion, education, and research come full circle at Omaha's Henry Doorly Zoo and Aquarium. Creating a truly great experience I have had the privilege to document, you can be sure that I will go back again to see what other amazing things the future may bring.

Western Lowland Gorilla
Status: Critically Endangered

Omaha Biome

An Open Heart
Bleeding-heart Dove
Status: Near Threatened

Feathered Mohawk
White Crested Laughing Thrush
Status: Least Concern

Made for This
Spotted-necked Otter
Status: Near Threatened

Tailless Grip
White-handed Gibbon
Status: Data Deficient

Jungle Portrait Mode
Malayan Tapir
Status: Endangered

Furry Oxymoron
Giant Elephant Shrew
Status: Least Concern

Miniature Gaggle
African Pygmy Goose
Status: Least Concern

Can I Help You?
Indian Flying Fox
Status: Least Concern

Handy Grip

Snack Attack
Egyptian Fruit Bat
Status: Least Concern

Full Belly

Cascade Snoot
Baird's Tapir
Status: Endangered

New World Soul
Black-headed Spider Monkey
Status: Critically Endangered

Sticky Fingers
Mexican Leaf Frog
Status: Least Concern

Before the Dawn
Magnificent Tree Frog
Status: Least Concern

Desert Snow Globe

Desert Rose
Klipspringer
Status: Data Deficient

Owl Doppelganger
Tawny Frogmouth
Status: Data Deficient

Getting Cleaned Up
Hottentot Chestnut Teal
Status: Data Deficient

Cooling Nook
Yellow-footed Rock-wallaby
Status: Near Threatened

Power Nap
Coatimundi
Status: Least Concern

Red Hot Bite
Black Mamba
Status: Data Deficient

Stalactite Gateway

Oddball in the Night
Aardvark
Status: Least Concern

Reversed Cycle Hopper
Springhaas
Status: Least Concern

Double Vision Leucism
American Alligator
Status: Least Concern

Acting Judge
American Alligator
Status: Least Concern

Flowing Locks
Angolan Colobus
Status: Least Concern

Human

Chimpanzee

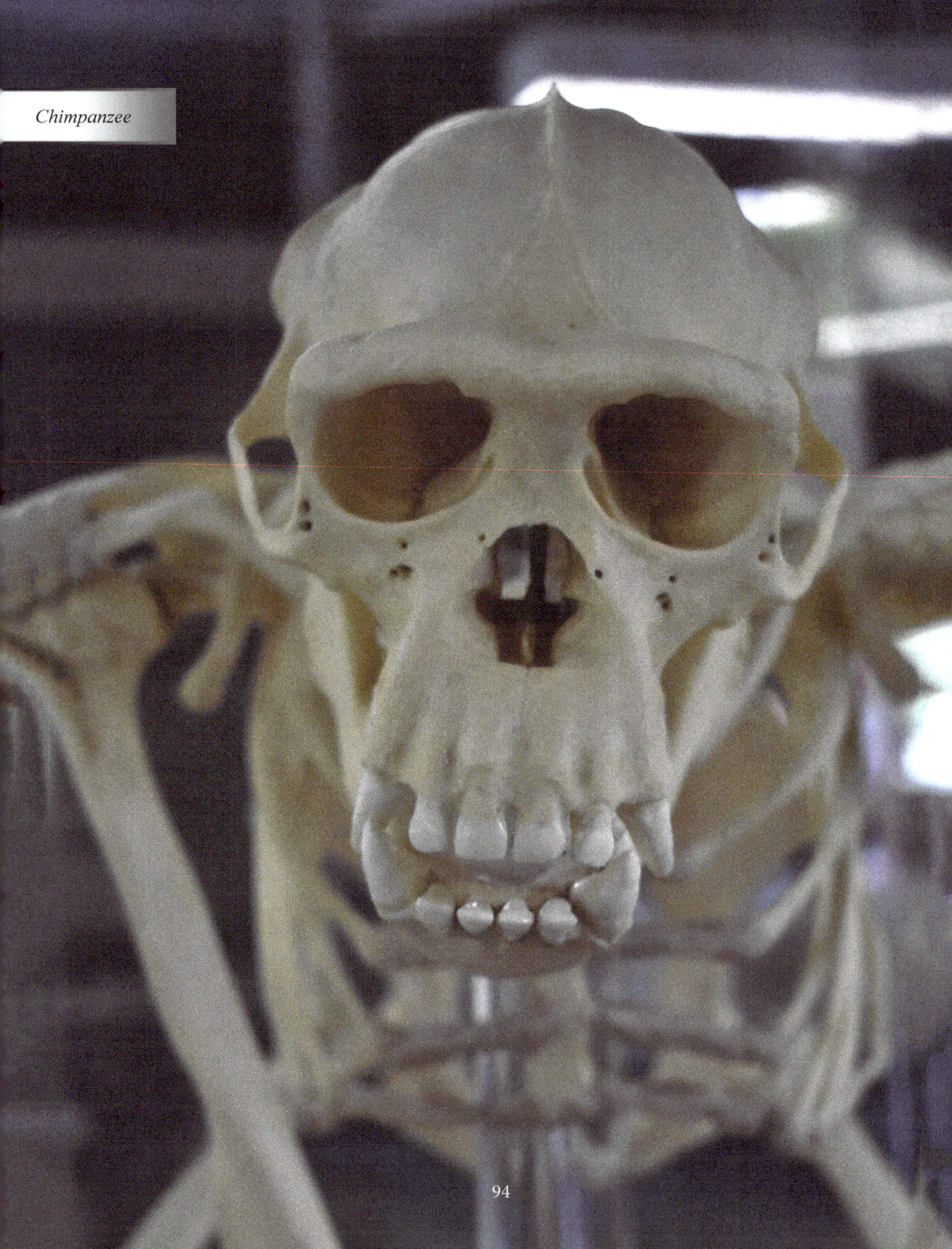

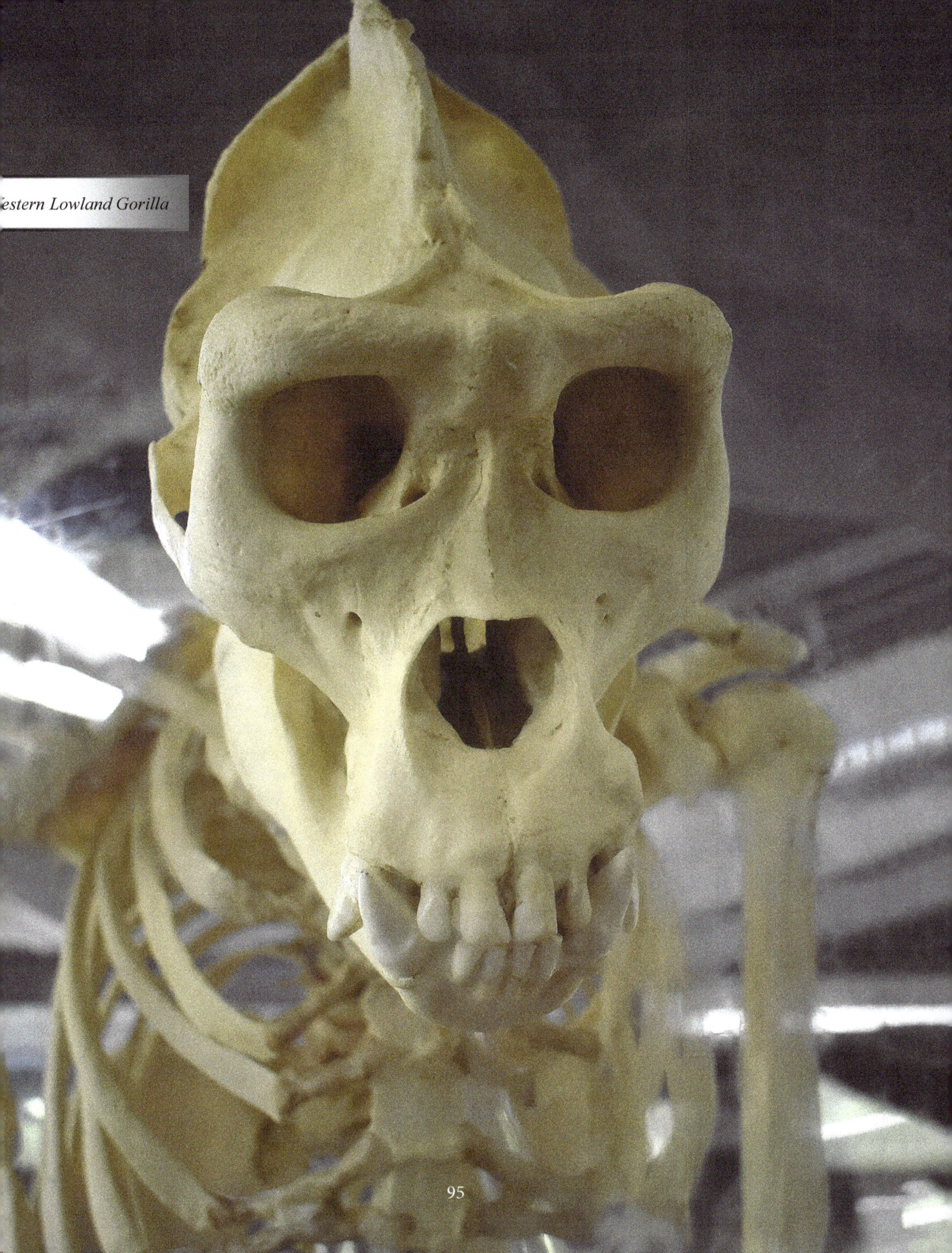
Western Lowland Gorilla

Orangutan

Early Development

Stern Alpha
Western Lowland Gorilla
Status: Critically Endangered

Vanishing Gem
Mongoose Lemur
Status: Critically Endangered

Freshman Sprinter
Cheetah
Status: Vulnerable

Blue Moon Beauty
Moon Jelly
Status: Not Evaluated

Pacific Mob
West Coast Sea Nettle
Status: Not Evaluated

Toronto Zoo

Change Of Season

As I continued this project, I aimed to broaden my horizons and go outside of my comfort zone. After a very busy summer I was ready to go beyond the borders of the United States and make my first stop at the largest zoo in all of Canada, The Toronto Zoo. I decided to go to this facility only knowing the basics of where it is and when it opened.

Little did I know that leaving a glimmer of mystery out of my research would lead to the discovery of some lovely surprises. Making every second of my tour better than I could've hoped for.

Trust me when I say that any little moment of good luck and levity were much needed considering a few snags that I encountered. To start off, the forecast was against me the whole time. Three days full of clouds, cold temperatures, and constant drizzle. Secondly, as I awoke on day one of the trip I was blessed with a burning sore throat and horrid cough. In times like these you come to find that little obstacles will have a habit of testing your durability and passion.

All I knew was that, come hell or high water, I was going to explore all 700 acres that hosts nearly 5,000 animal residents by the time the trip was over.

Toronto Zoo brilliantly sowed each exhibit into the natural environment that Ontario had to offer. Those very environments continue beyond the perimeter of this facility with the Rouge National Urban Park, making this zoo the only one in the world to be at the gateway of a national park. This unique affiliation allows for a beautiful relationship between nature and the public that can continue after leaving the zoo.

Early on, I came to find a phenomenally high presence of educators and keeper talks around the clock at the zoo. Even during the dreary weather in the off-season, you seldom would find yourself without a caring knowledgeable staff member nearby willing to share and teach. Toronto Zoo actually has approximately 400 volunteers dedicated to education, and in 2017 they logged over 29,000 hours of service! They all take time out of their lives to spread positive messages of conservation and knowledge. Each zookeeper and volunteer I talked to were so engaging that whenever I would ask questions or show intrigue, their faces would light up. Those people-to-people connections are what can inspire a future biologist or even have someone change one habit to help the earth. Seeing all that infectious positivity from the staff gave me energy to forget all my little turmoils of the trip and focus on showing the beauty of every species I could.

Several other moments of sheer joy were when I visited the Pygmy Hippos, Greater One-horned, and Southern White Rhinoceros. Each of them had an amazing healthy calf, ranging from two months old (Pygmy Hippo) to about a year old (both rhino species). To be present, as the next generation is starting to explore, is truly a privilege. Their natural curiosities and play behaviors are all a critical part of their development. Each one of these great creatures only have one calf at a time, long gestation periods, and stay alongside their mothers for an extended period of time. These factors combined with the dangers facing their wild counterparts (poaching/encroachment) makes every birth a precious addition towards the success of the species.

I almost had completely missed my chance to see the Southern White Rhinoceros. Due to the weather, they had access to their indoor living spaces for whenever they wanted. So I would keep making my way across the zoo just in case they were outside, but each time they chose to take a break from the weather and stay cozy. Until…on my very last day in the country with two hours left before the zoo was to close, I decided to try one last glimpse. I arrived at their immense plains exhibit to find the whole "crash" (group of rhinos) just wandering out. As the adults were taking advantage of the fresh mud, the little one was practicing sparring. The calf then began a nice easy walk-about and, within a minute, turned it into a full on playful sprint. His mother was soon taking his lead and before I knew it, they were running side by side. These wonderful beasts were running full speed with space to spare creating a truly majestic moment. One of many that occurred on this trip.

Speaking of majestic, in the previous chapter I had mentioned how leucism can occur in multiple species. One of those very possible species is the African Lion. I have never witnessed this variation in lions in person. So just think to yourself how one would react upon stumbling across this kind of creature. I stopped in my tracks with my head turning like an owl saying out loud, "Is this real?!" White Lions were once seen frequently in parts of South Africa around the Timbavati region, but as the time went on, trophy hunters would flood the area hoping to bag a rare prize. This in turn caused the rapid decline of these unique cats. The expression of this trait is believed to be expressed at higher frequency in this region due to the light sandy soil that is prevalent there. Over the generations this trait would greatly assist them in the camouflage game while hunting their prey. With less than around 300 individuals in captivity and only rare isolated pop-up sightings in the wild has this gem facing an uphill battle.

Not only has the Toronto Zoo committed to many conservation programs at home and abroad, but they have also made some impressive green strives at their own facility. To me the standouts were the ones that could be seen and experienced by all the guests at the zoo. These included modifications to a couple of their buildings with green roofs and the use of geothermal heating/cooling. Another fantastic project lies with the future use of poo at the zoo.

Installing green roofs on existing buildings is a bright part to the future of sustainability and saving energy. The roof is given extra layers for support, one for drainage, a growth medium, and then capped off with different vegetation. These were implemented on top of the Australasia pavilion and Tundra exhibit. The extra growth on top of just one building reduces carbon dioxide and can produce a single year's worth of oxygen that's required for one person. The buildings also maintain a cooler temperature, experience reduced noise pollution and energy consumption. In recent studies it has been discovered that if a mere six percent of Toronto's buildings were converted to green roofs, there would be an estimated $1,000,000 in energy cost savings.

Geothermal temperature regulation is a practice where the natural energy from the earth is utilized in order to keep a certain area warm in the winter and cool during the summer months. This form of space heating has proven to have the lowest environmental impact compared to other technologies in the field. The zoo has taken the initiative to install this technology into their Lion-tailed Macaque exhibit. Guests can experience the positive effects of this technology in the winter as they witness these primates in their outside areas in the winter with snow melted away from most of the area in the exhibit. There is also an area by the guest paths where you can feel it working if you put your hands down toward the mulch.

Finally, the plan that has been underway, and hopefully opens in May of 2019, will be the first zoo based biogas plant in North America. Old food from a supermarket chain and animal waste from the zoo will be processed having the biogas power the zoo. The power generated will account for one-third of the zoo's needed energy. Greenhouse emissions would also be greatly reduced by nearly 10,000 tonnes in a single year! Every advancement we pursue can make a very significant impact to our world down the road. In my eyes, Toronto Zoo is certainly leading a charge toward a bright future.

Counting Sheep
Arctic Wolf
Status: Least Concern

Downward Dog

Northern Giants
Wood Bison
Status: Near Threatened

Rorschach Cat
Clouded Leopard
Status: Vulnerabl

Little Blue Bell
Blue Poison Dart Frog
Status: Vulnerabl

Rorschach Cat
Green and Black Poison Dart Frog
Status: Least concern

Double Trouble
Yellow-banded Poison Dart Frog
Status: Least Concern

Coiled Patience
Green Tree Python
Status: Least Concern

I've Been Spotted
Fairy Bluebird
Status: Least Concern

Leap of Faith
Matschie's Tree Kangaroo
Status: Endangered

Out of Steam

Work Smarter, Not Harder
Sumatran Orangutan
Status: Critically Endangered

Ghost Cat
Snow Leopard
Status: Vulnerable

Cheers To life

A Hopeful Future
Greater One-horned Rhinoceros
Status: Vulnerable

Swimming Mosaic
Mbuna Cichlids
Status: Varies among these 19 species

Brilliant Stand-Out

Exposed Spine
Nile Soft-shell Turtle
Status: Vulnerable

Moonlight Grazers
River Hippopotamus
Status: Vulnerable

Pygmy Hippopotamus
Status: Endangered

Time For Bed

Little Morningstar

El Rey Blanco
White African Lion

Queen Of Stone

Sparring Partner

Tandem Sprint
Southern White Rhinoceros
Status: Near Threatened

ZSL London Zoo

Powerful Message
ZSL London Zoo

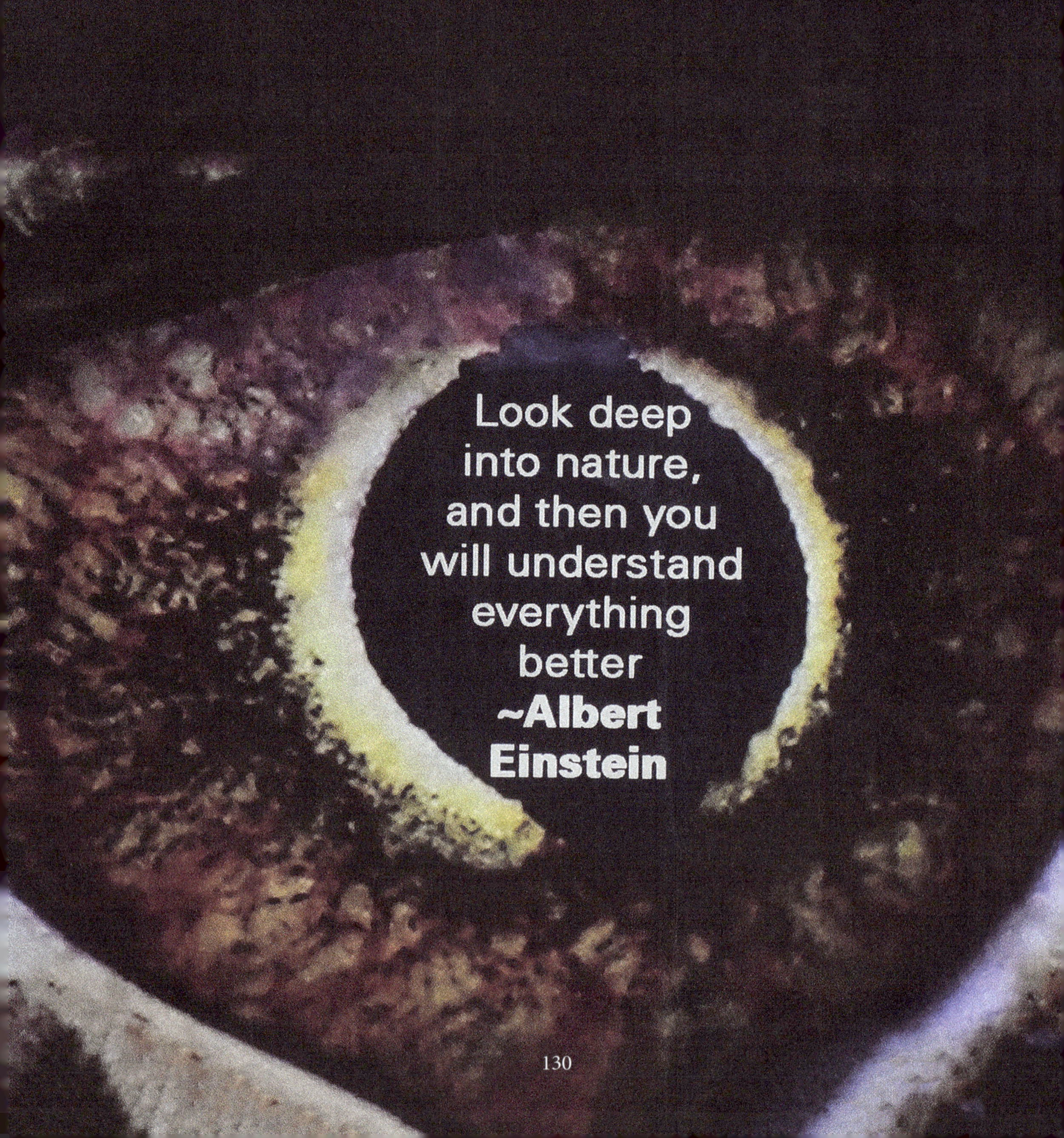

Look deep into nature, and then you will understand everything better
~Albert Einstein

With the year coming to a close and being filled to the brim with spoils from a Thanksgiving feast, it was finally time to journey across the pond. The next stop on my quest was the ZSL London Zoo in England. Of course, with this being my first trip overseas in almost a decade, it wouldn't be complete without the accompaniment of anxiety driven plane trouble. Even though boarding the plane was flawless, a luggage snafu turned what would've been a three hour layover in Iceland into a significantly shorter five minute connection. Once again, lady luck was on my side when I landed. The gate for my connecting flight was straight across the corridor. After a brisk run I made it on the last shuttle taking passengers to the plane out on the tarmac. Phew! With that out of the way, I was more than ready to explore the world's oldest scientific zoo.

Before we dive right into this trip, let me give you a little background on what the ZSL actually is. The Zoological Society of London (ZSL) was established by Sir Stamford Raffles in 1826. It serves as an international scientific, conservation, and educational charity. Their goal is to globally enact crucial conservative efforts for countless species and their corresponding environments. In 1828 early scientists of the ZSL opened the zoological gardens where they could come together in order to study and share knowledge for the scientific community. By 1846 the garden was host to the most extensive collection of animals in captivity and was opened to the public that same year. Considering the travel capabilities of that time period, Londoners truly could see the gems of the world that many people had no idea may have existed. From then on, the gardens became known as the ZSL London Zoo, and thus continued a rich history.

I was now walking the same grounds that Sir Charles Robert Darwin did when he pieced together his findings from his ventures. With the help of ZSL London Zoo scientists, he identified finches from the Galapagos. He also observed the Orangutan there and felt that this was an obvious factor between ape and human relations. Another well-known person who frequented this institution was the author A.A. Milne with his son, Christopher Robin. Here, he found inspiration for the renowned works many people have grown to love, thanks to an amazing Black Bear named "Winnie." This facility is also the site of the world's first reptile house (1849) and the first public aquarium (1853).
Both of these historic exhibit houses, as well as the rest of the zoo, hold important messages and action plans in place to conserve those who need
the most help.

I have always loved to get up-close with all kinds of reptiles because of how important they are with keeping the jigsaw puzzle of nature together. The reptile house was filled with an amazingly diverse educational collection, along with ones that I have never seen before. A few of those species were even discovered by ZSL scientists. The exhibits were not only wonderfully suited for the herps, but beautifully conveyed the plights which these creatures face. One of the biggest troubles is the illegal wildlife trade. This nearly seventeen billion dollar tragic industry has led many animals to the brink of becoming extinct. The Annam Leaf Turtle is one of the many species that has suffered because of that industry. However, ZSL London Zoo has been able to help them take a step in the right direction by working closely with local governments in their endemic regions.

As I continued through the reptile house, I learned about a program that is enacted throughout the rest of the facility that really ignited my interest. It is called the "EDGE" program. The ZSL have identified species that are Evolutionarily Distinct and Globally Endangered using the latest techniques from the scientific community. Those that are part of the program have few close relatives and are peculiar both anatomically and in terms of behavior. Since most of these creatures are not very well known, they slip under the radar of awareness to the everyday public. Every person should be able to relate to animals like these. Uniqueness and peculiarities are what sets every person apart, we all hold something different and those things make us special. That is why the EDGE program spoke to me as something phenomenal.

Next, I took a short jaunt right across the path to see the other historic landmarks in ZSL London Zoo's history, the world's first public aquarium. Believe it or not, the word "aquarium" was created by the ZSL by abbreviating the term "aquatic vivarium." As the aquarium's popularity rose with the public, it was rebuilt under the zoo's Mappin Terraces in 1921, allowing additional space for more exhibits and educational opportunities. One such opportunity is at the coral nursery exhibit. The nursery is composed of rare corals that were confiscated by UK officials. Once they reach a certain point, the corals are used both here and other European aquariums to help keep populations fresh. Coral is an extremely vital part of the oceans and acts as an environmental indicator. When populations of coral are stable and healthy, other life around will flourish as well. So with that, it was now time to finish the rest of this zoo tour.

Going through the remaining thirty-six acres of this zoological institution, it was so easy to forget that you are smack dab in the middle of a major world city. Especially, as you gaze upon the splendor of the Penguin Beach and Tiger Territory exhibits. At Penguin Beach, the 450,000 liter pool allows their resident Humboldt Penguins to fully engage their evolutionary pristine bodies. They move through the water at full speed and, just like a little kid, I wanted to keep up with them as they passed by me. At the far end of the exhibit lies a row of nesting sites where the penguins that have paired-off lay and incubate their future chicks. The environment allows them to maintain an amazing lifestyle and propels a great breeding program.

Sadly, in early 2019, their resident female Siberian Tiger, Melati, had passed away. Even though I had only spent a couple days at the zoo, the moment I heard the news it did strike me emotionally. The ZSL London Zoo had managed to build a connection between Melati and me as the guest. Everything from the educators telling me about the predicaments wild tigers face to watching her explore her space and be a true cat. When an unfortunate thing like this happens it is never easy. As animal care professionals, we build strong bonds with those we care for. We give them the best life so we can share their stories and why they matter in this world to the public every day. Every creature is special. They all play a role in the big picture we call life. It starts with us. We need to do what we can to keep the lives of those in nature going.
That is the message that ZSL London Zoo conveys and spreads to all who visit their institution.

By Land and Sea
Caiman Lizard
Status: Least Concern

Suspended Animation
Lake Oku Clawed Frog
Status: Critically Endangered

Dive Master

Dive Master
Dumeril's Salamander
Status: Critically Endangered

Wandering King
King Cobra
Status: Vulnerabl

Self-Discovery
Mindanao Water Monitor
Status: Least Concern

Face to Face
Indian Spotted Pond Turtle
Status: Endangered

Find Your Happy
Annam Leaf Turtle
Status: Critically Endangered

Generating a Future
Coral Nursery
Status: Varies among species

Master Diver
Humboldt Penguin
Status: Vulnerable

No Tourists Here

No Shave Required
Bearded Pig
Status: Vulnerable

Smells For Days

Facing the Next Generation
Giant Katydid
Status: Not Evaluated

Ring the Dinner Bells
Golden Silk Orb Weaver
Status: Vulnerable

Found the Seeker

Living Foliage
Southern Tamandua
Status: Least Concern

Sight Through the Darkness
Gray Slender Loris
Status: Least Concern

Miniature Wisdom
Emperor Tamarin
Status: Least Concern

You're Making Me Blush
Red-faced Spider Monkey
Status: Vulnerable

Naturally Handsome
Sumatran Tiger
Status: Critically Endangered

Unbroken Gaze

Insight to My Soul

In Loving Memory of Melati

ZSL Whipsnade Zoo

While preparing for my visit for to the ZSL London Zoo, I stumbled upon another landmark facility also located in England, ZSL Whipsnade Zoo. This institution, as you may have guessed, is also operated by the Zoological Society of London. With a packed schedule, only allowing for a one day adventure at ZSL Whipsnade, I was dead set on exploring every inch. Seeing as ZSL Whipsnade is the biggest zoo in the United Kingdom, sitting on 600 acres, I had some serious work ahead of me. The history here was also amazing just like its sister facility (ZSL London Zoo).

Starting with only a small collection of three pheasants and five red jungle fowl, the zoo has now grown to host around 3,500 brilliant creatures. In 1931, the ZSL Whipsnade Zoo opened its doors to the public, becoming the world's first open zoological park. Only a few years later, they became the evacuation site for the inhabitants of the ZSL London Zoo during WWII. Located about an hour north of the city, it was the ideal buffer to the effects of war.

While I made the trek to the zoo in Dunstable, every bit of the countryside was beyond gorgeous. The endless hills of green with a side of tranquility allowed me to relax and truly focus. This is not normally hard to do under regular circumstances, but most of the time I am never part of normal circumstances. As expected for that time of year in England it was dreary. The only difference here was that I wasn't the only out-of-town visitor; Storm Diana visited the island that very same day. Of course it was nothing crazy… just mere 20-40 miles per hour wind gusts and constant rains for the course of my visit. I'll be the first to admit I was not prepared for the weather. A light windbreaker and jeans did not keep me dry by any means. On the bright side, I had the zoo all to myself, my gear stayed dry, and a majority of the animals were out and about.

Along with great opportunities to capture amazing moments, there also were shining examples of extraordinary conservation and educational programs. Some of these included the African Painted Dogs, Scimitar-horned Oryx, and Asian Elephants. Each have faced challenges that led the ZSL to take the wheel and steer them towards a positive outcome over the years.

African Painted Dogs are some of the most successful hunters on the plains of Africa. In 2017, ZSL Whipsnade Zoo opened a new exhibit for them that encompass 8,500 square meters. When I arrived, the pack was relaxed in a cuddle puddle waiting out the weather in their cozy indoor housing. I have found that many people who stumble upon animals resting tend to become disappointed or even upset. In all my years, I have never had those feelings simply because I know all creatures need rest. Many carnivores sleep for extended periods of time, it's in their biology. An animal resting peacefully will also be an indication they are comfortably content in the present environment. While reading the bountiful educational materials at the exhibit, the pack slowly awoke, setting out to patrol their territory.

Watching them follow their leader and fanning out was like having a front row seat to a tactical military drill. In the wild, all their skill and precision in hunting can lead to conflict with humans.

The human populations that live on the plains of Africa rely heavily on livestock for their entire livelihoods. As the habitats between people and the African Painted Dogs crossover, the livestock find themselves sometimes becoming prey. In an effort to combat any negative conflicts, the ZSL is working with the locals in Kenya to establish a manageable coexistence. By sharing scientific research and providing a continued fluid line of education to the public, the amount of conflicts have reduced while the Painted Dog's numbers have begun to rise. Another formidable threat in the way of this species has been rabies. The ZSL have partnered with the Mpala Research Centre, reaching out to the communities willing to vaccinate for this virus in their domestic dogs.

As the world has become smaller, I am sure you have realized that human conflict is the primary source for many species facing a decline. In my opinion, the Scimitar-horned Oryx's plight was one of the most devastating examples.

Once holding prevailing numbers across the deserts of Chad and Niger, they quickly succumbed to overhunting due to the civil turmoil that rocked the region in the 1980s. By the change of the millennia, this species was declared extinct in the wild by the IUCN. Luckily, there was a glimmer of hope. This Oryx species still existed in over 220 zoological facilities and private institutions across the globe. The ZSL was not alone in wanting to keep them from completely disappearing. So they decided to become a part of a massive undertaking with the end goal of reintroducing this species to their endemic lands. It was found that the Ouadi

Rimé-Ouadi Achim Faunal Reserve in Chad would be the ideal spot for a reintroduction program. Alongside the ZSL, (and other partners), the Government of Chad, and Environmental Agency of Abu Dhabi (EAD) hatched a plan.

The EAD hosted a herd of Scimitar-horned Oryx that were pieced together from institutions around the world, including two members from ZSL Whipsnade Zoo. These individuals were selected in order to promote the greatest genetically diverse population. In March of 2016, the first group was set fourth into the reserve. Each member was outfitted with GPS tracking collars and was closely monitored by members of the Smithsonian Institute. By September of that same year the first calf was born. This was believed to be the first wild birth for a member of this species in nearly three decades!

Later in 2017, another herd composed of fourteen members was released into the same area. These small yet significant triumphs showed us all that multiple organizations and governments can come together and accomplish something.

The last exhibit I visited provided amazing educational insight to any of the guests who may cross its path. The Asian Elephant Centre for Elephant Care was opened by the Queen of England in 2017 and is by far one of the most amazing elephant barns I have ever seen. Composed of soft sand floors along with timed feeding systems provide the residents with a naturalistic varied environment around the clock, even if they are not able to access their outside paddock. They also have an ample amount of displays which give amazing facts to help intrigue the guests about the lives of elephants everywhere. With nearly 20,000 elephants killed around the world in 2015 due to the demand for ivory, every addition to the species is dear to the world. Even with all the wind and rain, the two young calves that call ZSL Whipsnade Zoo home were building their strength and learning how to gain control of their trunks through vigorous play. Watching the development of these gentle giants is yet another element that will never grow old to me.

As I ended my day wandering towards the exit, I couldn't help being filled with an incredible sense of appreciation for all of the creatures around us. For someone who has always had a passion for wildlife, I didn't think I could appreciate them more than I already did. ZSL Whipsnade Zoo was a gem of a surprise that I stumbled upon while researching their sister facility. All the bad weather and a long journey there from London were well worth all the experiences and knowledge that I get to share with you.

Sure Footed
Eurasian Lynx
Status: Least Concern

The Face of Adaptation
Wild Boar
Status: Least Concern

Resting Warrior
Wolverine
Status: Least Concern

Grab and Go Fishin
Asian Small Clawed Otter
Status: Vulnerable

Top Dog in Pasture
Gaur
Status: Vulnerable

Ohhhh, I Smell It

Leaving Leaf

Climbing with Care
Red Panda
Status: Endangered

True Survivor
Scimitar-horned Oryx
Status: Extinct In The Wild

Built to Be Tough
Bactrian Camel
Status: Critically Endangered

Herd Of Hills

Painted Predator

Nimble And Swift
African Wild Dog
Status: Endangered

First Patrol

Barn Innovation
Asian Elephant
Status: Endangered

Head Strong

Tree Trunk Trial

Pure Joy

The End of the Wandering Road

Artwork: Gaia
Artist: Luke Jerram

Artwork: Gaia
Artist: Luke Jerram

Well, here we are. After a year of traveling while maintaining a full time job, I find myself on my last adventure. I took a chance and reached out to an amazing zoological facility in the South Pacific to inquire if I could take photographs to be included in this book. They very generously granted me the permissions to photograph and I am still extremely thankful for the opportunity. For all those wondering which facility this was, you are just going to have to enjoy the pictures and not mind the anonymity. Per the agreement with this institution, their name will not be referenced.

But, let me assure you, this last stop was a truly marvelous end to an incredible year. As I wrap-up this wanderlust exploit, I hope you enjoy these photographs as much as I do.

This location allowed me to showcase some of the most unique, specialized, and deadliest creatures on this planet.

Even though this was the farthest I have ever been from my home, it sure didn't feel that way. Everywhere I visited I knew that I would only have one chance to capture the beauty of the amazing creatures they care for. Creating a connection to my audience and these zoological institutions has always been the main goal. I fully understand that many people will not get the opportunity to visit the same places I have in this book. All of these species and conservation programs were truly fantastic and I am happily compelled to share them in all their glory.

In regards to the conservation programs that all these facilities participate in, I have only shown you just a small preview. All animal care professionals in these locations not only have a relentless passion for wildlife, but continuously strive to create new programs. Any step made in the right direction when made by many can trigger significant change.

Around the globe, little by little, people have realized that in order for change to occur, there must be cooperation across all levels. To me, the most stunning example of this has been clearly seen in the city of Singapore. Over the course of five decades, an astonishing 2,000,000 plus trees have been planted around the city and they successfully cleaned up the waterways. They also erected several gorgeous fifty meter high tree-like structures that have every inch covered by creeper plants along with other vegetation. All of these changes to the city have brought bustling amounts of animal life to the urban environment. It turns out that Singapore has become host to the richest amount of species over any other major city in the world.

Two other tools that can play a vital role in helping a species bounce back from dangerous low numbers is education and raising awareness. A prime example of this was seen on the island of Trinidad with Leatherback Sea Turtles. Leatherback Sea Turtles are the largest species of sea turtles and often use this location as a nesting site for future generations. However, when they arrive on the beaches, their massive bodies present them with a slow struggle until they return to the water. These moments of vulnerability make them easy targets by locals harvesting them for meat. It led to a dramatic visible decrease of these majestic giants frequenting the shores.

A few members of the local population decided to spearhead efforts in order to reboot the turtle population while also finding a way to bring benefit to those people trying to sustain a way of life on the island. They showed their peers that tourism can generate a much better way of life than poaching. So now, many host-guided tours take place on the beach for visitors who want to catch a much desired glimpse of the Leatherbacks. Other members helping this species will excavate nests that were laid too close to the water to spots higher on the beach so they don't get washed away during high tide. Over twenty years, the combined efforts mentioned above have attributed to nearly a tenfold increase in populations across those beaches.

I am sure that by now you have gathered that all things take time. Time has shown what we have recovered, but it has also shown what we have lost. Over the last half a century, the populations of creatures on land and sea have decreased by nearly sixty percent. Over the course of a single year, an estimated 8,000,000 tons of plastic establish a residence in the oceans. Staggering numbers of species we see go extinct in our lifetime can possibly continue to rise, and the South Pacific floating garbage patch is likely to continue, if not met, with an opposing energy of change.

Yes, realities like these are hard to ignore. I am not telling you to immediately drop everything and get out there this very minute. I am just trying to show you the realistic challenges that our nature is facing. We can all do many little things to help them out. Properly recycling, taking a few hours to pick up rubbish on your local beaches, or even taking the time to research the everyday products that you use are made sustainably. Also, if you would like to join me by visiting zoological institutions like the ones I've mentioned in this book, I would not mind the company.

Like I've said before, animal care professionals across the scientific communities have a drive to protect wildlife. Wandering around our planet has truly been an unparalleled experience that will not be forgotten and taken for granted. All life is precious. Showing just a small sample of such life to all of you through my photography has become a growing passion I hope to continue. I always hope to be a witness in the lives of these creatures, to display all the beauty and importance they possess. Once again, thank you all very much for joining me. May the wanderlust ideal always push our curiosities, drive our thirst for knowledge, and explore the limits beyond our comfort zones.

Not for a Normal Thanksgiving
Brush Turkey
Status: Least Concern

*Beautiful Camo
Eclectus Parrot
Status: Least Concern*

Always Curious
Bush Stone-curlew
Status: Vulnerable

Feathered Judgment
Wedge-tailed Eagle
Status: Least Concern

Striking Grip

Fluorescence in Cover
Jabiru
Status: Least Concern

Zebra Tail Portrait
Ring-tailed Lemur
Status: Endangered

Wise Beyond His Years
Binturong
Status: Vulnerable

Favorite Spot

Room to Grow
Rothschild Giraffe
Status: Near Threatened

First to Feel Rain

Changing of the Guard
Meerkat
Status: Least Concern

Fluffy Tank
Southern Hairy-nosed Wombat
Status: Near Threatened

Riverbed Cushion
SAgile Wallaby
Status: Least Concern

How Does My Fur Look?
Swamp Wallaby
Status: Least Concern

Every Last Bite
Short-beaked Echidna
Status: Least Concern

Spiky Toupee

Compact Powerhouse
Tasmanian Devil
Status: Endangered

Accented Howls
Alpine Dingo
Status: Endangered

Always Watching
Fierce Snake
Status: Least Concern

Always Ready
Tiger Snake
Status: Least Concern

Baby Steps
Northern Koala
Status: Vulnerable

Eucalyptus Eclipse

Anchored Resting Spot

Pure Love

Patient Predator
Freshwater Crocodile
Status: Least Concern

Sensory Overload

Prehistoric Present

The Snap Above All Snaps
Saltwater Crocodile
Status: Least Concern

One Final Look

One Final Look

A Twist of Fate

Bibliography

Blue Planet II. Perf. Sir David Attenborough. BBC Natural History Unit, 2018.

The IUCN Red List of Threatened Species. Version 2019-1. www.iucnredlist.org.

North Carolina Zoo. May, 2018. www.nczoo.org.

OdySea Aquarium™. February, 2018/2019.

www.odyseaaquarium.com

Omaha Henry Doorly Zoo and Aquarium. May, 2018. www.omahazoo.com

Our Planet. Perf. Sir David Attenborough. Silverback Films, 2019.
Planet Earth II. Perf. Sir David Attenborough. BBC Natural History Unit, 2017.

Toronto Zoo. October, 2018. www.torontozoo.com

Zoological Society of London: ZSL London Zoo and ZSL

Whipsnade Zoo. November, 2018. www.zsl.org

www.ingramcontent.com/pod-product-compliance
Lightning Source LLC
Chambersburg PA
CBHW041919180526
45172CB00013B/1331